What I've Done

Robert F. Baudendistel

authorHOUSE®

AuthorHouse™
1663 Liberty Drive
Bloomington, IN 47403
www.authorhouse.com
Phone: 1-800-839-8640

Published by AuthorHouse   10/25/2012

ISBN: 9781477248010 (sc)
ISBN: 9781477248003 (e)

Library of Congress Control Number: 2012912801

Preface

THE INTENTIONS OF THIS book are to give the reader some knowledge about what jobs I held on a part-time basis to support my family and remain in teaching.

I initially worked for others in various capacities before I realized that I could be self-employed and have the monies all come to me.

I was well versed in the fields of horticulture, masonry and carpentry.

The text of this book also envelopes the field of education and its commitment to the growth of young adults.

The book also touches on the camping experiences of physically handicapped children and their ability to adjust to a new environment.

The book also acknowledges some of the ills of our society that still persist to this day with suggestions as how to correct them.

Contents

Radiation

MANY EXPERIMENTS HAVE BEEN conducted on rodents with some form of radioactive material. They received, usually by injection, a solution containing a marked chemical designated to locate the area of the body where it was assumed to be deposited, i.e. the liver.

Simple detection of radioactivity can be performed by using a Geiger counter by placing it close to the object in question. To assess a more sophisticated reading, a spectrograph is used by plotting the results on a graph.

Every person should be aware of the dangers associated with radiation because of its invisible nature.

Experiments like this one are performed throughout the chemical industry on a daily basis.

The technicians have been taught to be extremely careful in disposing of the contaminated waste. They must develop a "mind set" of how to make everything aseptic by placing contaminates in air tight metal drums. Once accomplished, they are satisfied that they have protected their fellowmen from the ills of radiation. This is not always true. The person who has a license to dispose of nuclear waste may not be fulfilling his obligation.

Who follows the individual, paid by the chemical facility or has been cleared by the government to be pure and above approach? How many airtight containers have experienced a "sea burial" or have been left on a loading dock somewhere?

Oak Tree

MY TASK FOR A particular Saturday evening was to deliver a completely equipped boat and trailer to its new owner. This plan was orchestrated by Pierre, the boat shop owner to be ready anytime on Sunday.

The only criterion left undone was to load the boat with food and drinks for their full day excursion. On the way home, a 30 mile trip, the weather became nasty with fog and a light mist causing poor visibility. Acknowledging this traveling condition, I reduced my speed which caused the rear of the truck to sway and fishtail.

I corrected the sway but noticed the VW Beatle coming towards me, attempting to pass a car on a curve.

I eased the truck (away) to my right to avoid an accident when suddenly the truck turned 180 degrees and began going backwards faster than I had been going forward but still on the same side of the road.

I was convinced that I was going to meet my maker at this time and place in the final seconds of my life. I stopped abruptly because I had just collided with a massive oak tree. The middle section of the truck's tailgate experienced the most damage. The installed heavy duty wench made the damage to the truck less severe.

I do not believe that the old adage "In a moment of peril, your entire life passes before you" was what I had experienced. I remained calm yet aware of how serious it could have been.

Baseball

I HAD JUST MOVED into a new neighborhood when I received a phone call from a member of the community. He wanted me to coach a little league baseball team (ages 9-11). I told him that I had played baseball all my life but had never been involved in coaching the sport. I asked how he received my name. He told me that my name had come up numerous times in his dealings with the community.

I agreed to coach the team and received 18 names as members of the team. These boys had been selected by a potential coach two months before they were engaged in practice. He was transferred by his company's main office, therefore his coaching career never got off the ground.

Even though I had enough boys to make two teams, I experienced extreme difficulty in having more than ten show up for a practice or a game. To be successful and make certain that I could field a team, I elected to pick them up myself, whether it was a practice or a game.

I was not very popular with the mothers for holding practice on Mother's Day but received thanks for holding practice that particular day from the fathers. On that day, most of the fathers joined in practice with their sons.

It is my firm belief that baseball will not be meaningful for any person, male or female, who has difficulty hitting the ball.

Here are some comments that I feel should be made:

1. Batting practice should be limited to using one ball at a time, especially for little leaguers. The procedure is much slower but is much safer.

2. Maintain order at all times and tell parent and family members not to be confrontational with others while at the ballpark.

3. The rationale that smaller players will help their team more by taking a walk rather than making an attempt to make contact with the ball is wrong. The request "Don't swing, let him give you a walk" should be ignored if our aim is to produce baseball players.

4. The little league handbook states that every player is to be placed in the field for 1–2 innings and record a minimum of three outs. This handbook should state that every player must be given one chance to perform as a batter. Too often the coach is able to manipulate his scorebook to decrease the possibility of having to use one of his less talented players. The most common method employed by coaches of little league is to make the substitutions in the lineup in the sixth or final inning. If the substitute enters the game in the final inning as a member of either team, he may never come to bat.

Cap Rock

Murphy's Law states that "whatever is conditioned to go wrong will go wrong". This concept is readily apparent, while other times it is not.

Take this case in point. I was called by a "sister" to see if I would be interested and had the talent to reposition a large cap rock that had been uplifted from its original horizontal position. It came to rest on the stairs it was designed to protect.

This one piece of stone measured 6 'by 10' by 6 inches. The sister was elated when I told her I could reposition the stone. I informed the sister and my crew how I intended to put it back in place. We would employ a 2" x 8" x 12' joist to serve as a ramp. The joist would be placed lengthwise on the steps to accept the stone. We would slowly, yet carefully, slide the stone up the joist. One crew member would be positioned on each side of the front portion, there mainly to guide the stone as we slid it up the joist. The other two members of the crew pushed the base of the stone upwards and at the same time slid the joist upwards.

Every step that we traversed was easier than the one before. Once we had cleared the steps we placed the stone on the lawn.

Mortar was then mixed and distributed over the entire surface of the area designed to receive the cap rock.

The task became difficult again because positioning the rock could only be done by two crew members. One more problem we had to encounter was to free my middle finger. It had become wedged in the joint between the two cap rocks.

The stone had been hit by a drunk driver whose car had been inadvertently engaged to move forward, leaving the confines of the parking lot and heading towards the nun's living quarters.

It was fortunate that no one was using the stairs at this point in time.

Cabin

WE DECIDED TO FIND out why we had experienced excessive talking in the girl's cabin long after they had been sent to bed. I informed counselors, both male and female alike, that talking in their cabins long after dark would be a punishable offense.

I walked down to the girl's cabin from the dining hall. As I approached the cabin, I could hear laughter, singing and conversation. I was aware that in complete darkness, the campers were thoroughly enjoying themselves.

I did not turn the lights on as I entered the cabin. I wanted to see if I could assess what was happening in total darkness.

Lo and behold: Our blind camper, named Maria, was reading to her cabin mates a story from a book printed in Braille.

What could I say? I still become choked up every time I have a vision about that evening. What I had experienced was a pure expression of "love for your fellow man".

Landscape

MY FIRST DAY ON the job as a landscaper was varied and challenging. I was assigned to a three man crew.

Our first job was to plant two pyramid shaped evergreens on each side of the headstone. While driving to the gravesite, our foreman informed us that the records for the site had been destroyed by fire. The area was later filled in with top soil. The foreman laughed when he suggested that this area might contain some skeleton remains from years gone by.

I first made two circular impressions designed to mark the two holes. The foreman agreed with their location. His plan was simple. The newest member would man the pick, I the shovel, while the foreman would bring the plants and water as needed to the planting site.

The foreman went down the slopes, removed one of the trees and was proceeding up the hill when he called to me to explain why the pick user was lying on his back. Imbedded in the pick was a small portion of a plastic doll's head.

Seeing this object and remembering the earlier conversation was too much for him. He had passed out not knowing what had just transpired but was reawakened after being doused with water. He asked to be returned to the shop because he felt sick.

My foreman left me behind to plant the two trees while he took the sick man back to the shop. When he returned, I had finished the plantings, cleaned up the area and watered the plants.

Our next assignment was to prepare the soil for a foundation planting by spading up and turning over all the soil in the staked out area. To our amazement, the entire area beneath a thin layer of soil was deposited with round fieldstones. We collectively agreed all the stone had to be removed to a depth of three feet before planting could be done.

The foreman notified the owner of the dilemma. He told us to proceed with the stone removal. He would check with the home owner for a good location to pile the stones. This information had to be decided upon immediately for us to continue.

The house was part of a large tract of land being developed by the builder. To bury large amounts of stone below the surface might be common practice for some builders but it should be rejected. Imagine excavating for an in ground pool and have to explain to the home owner that an in-ground pool is not possible.

Car Transport

I HAD JUST BEGUN to use the ramp leaving the parkway when I saw a car swerve from side to side exiting in a cloud of dust.

I pulled my car off onto the road's shoulder, shut the engine off and ran up the sloping embankment to see if I could help the driver.

I was happy to see that both the driver and car had not been damaged. After several minutes of silence, the driver spoke. He said something was wrong with the steering. The car had begun to fishtail and sway like a baby carriage. On closer examination I noticed that part of the arm of the steering rod was missing a nut that was responsible for keeping the steering intact.

I offered to drive him to the nearest gas station to pick up a nut. The mechanic assured us that it had to be either one of two sizes and suggested we take both.

On the way back, the driver thanked me and expressed one more concern. He had to deliver the car before 5 p.m. to the Newburg boat ramp. He was assured a free bus ride back to Manhattan and $50 for transporting the car.

Little did I realize that I was involved in aiding and abetting a crime.

Months later a theft ring was uncovered. New cars were unlocked by using a slimjim. The ignition was compromised using a screwdriver. If the stolen car did not have sufficient fuel in the tank, the driver was told to fill it up at a "friendly" station. A false credit card was used at the station to keep gas records for each car separate.

The stolen cars were transported by an ocean ready tanker to South America.

Sunday Project

I RECEIVED A CALL from a woman over the weekend asking if I would be interested in restoring her home. I obtained directions and agreed to meet on Sunday at 10:00 a.m.

I arrived early and decided to look around to determine what had to be done. I could see from a quick examination that this small house had been moved to a higher elevation to avoid water damage by a nearby stream. The steel house beams were still intact and were supporting the foundation.

I decided that I was not interested in this project and started to leave. The woman appeared out of nowhere and acknowledged that I had been examining the correct house. I told her that restoring this house would not be financially sound. Tears welled up on her face followed by moans of frustration and she pleaded with me to help her, not deny her this wish. Every person she spoke to suggested that she cut her losses and forget restoration.

I told her that I was sensitive to her and that I would help her only on my conditions. In the following paragraph I enumerated these conditions.

This project was to be done only on Sundays during the hours of 7:30 am and 4:00 pm with a half hour for lunch (8 hour day). Adults would receive $100, teenagers $50. A weekly list of materials that would be needed would be ordered and charged to the owner's account. Two checks were to be written, one covering labor costs for the month just completed, the second for materials used during that same period.

One of the first orders of business was to remove the bottom three pieces of siding all around the perimeter of the house. It had been pried loose from the house frame but had never been stacked in an orderly pile.

I was turning one of the corners of the house when I stepped on a nail that was protruding through the siding. I expected to see the nail protruding through the top of my work boot since it hurt so much.

I worked the balance of the day, drove home and finally took my boot and sock off. The entire bottom of my foot was black. I decided to go to the hospital to have it checked out. The doctor told me to place my foot in a pail of Epsom salt solution. He asked how long had it been since I had a tetanus shot. Within the year I told him.

Months rolled by with success being attained at each phase of the project. Then the unthinkable happened, the owner was unable to continue the project because of lack of money. She did make one final request.

On the adjoining lot stood a new home for sale. She was perplexed as to why the house was still for sale. I pointed out to her that some of the items were constructed incorrectly. The kitchen pantry vinyl floor showed signs of stress before it had received any traffic. The joints of the subfloor had not been staggered to solve the problem.

The realtors in the surrounding area refused to list the home. They wanted the country bumpkins to beat the city slickers. They felt both insulted and abused by the new residents from Manhattan (the city slickers).

Instead of selecting a building foreman to make the job run more smoothly, they felt superior to the country bumpkins and could therefore determine how the house should be constructed.

Poolside

TWO WEEKS OF CAMP were over for the first group and the second group was scheduled to arrive that afternoon. A great number of good things had happened during the first session. Some of the campers arrived without a change in footwear and a change of clothing. Our aim was to seek donations from the radio listening public. The response to our plea was overwhelming.

I decided that it was a good time to rest and relax in my cabin during a quiet afternoon. All of the campers had been picked up by a family member after consuming the noon meal. Those remaining were the counselors, life guard, the kitchen help and the two county nurses all taking in the rays of sunshine at the edge of the swimming pool.

All of a sudden, I heard them shouting my name to come help get Bucky out of the pool. Bucky had bet a friend that he could swim 3 lengths of the pool underwater. He did accomplish the feat. He had previously consumed 3 spicy pickled sausages that had been sold from a large jar. Their consumption caused him to expel gas. When he released the gas from his stomach he took in pool water.

When I arrived at poolside, I saw his lifeless body lying on top of the pool's drain. Bucky had polio as a young boy which left his legs very small in diameter. His upper body was well developed, suggesting a well conditioned athlete. Charles, the lifeguard, with the assistance of one of the counselors brought him up to the side of the pool. I made a quick assessment of the situation and decided that because the pool was all concrete we had to place Bucky on a piece of plywood and lift him together with the plywood over the concrete side so that his legs were not damaged.

First the plywood, followed by Bucky's body was placed on the sidewalk that surrounded the pool. To me, Bucky seemed "blue and too far gone" to be saved by mouth to mouth resuscitation.

One of the nurses, Lee Ann, alerted us that Bucky was breathing on his own and was registering a heartbeat but was still unconscious.

The other nurse was on the phone requesting a state police escort to the nearest hospital (30 miles). They were told what roads we would be traveling so that they could intercept us. We would be in a gray Plymouth station wagon. We encountered three problems with our plan. 1). Top speed for the wagon was only 60 MPH. 2). other vehicles would only honor the police vehicle, 3). We were on our own once we entered the city limits.

Charles was happy that I called him a hero and that no one knew his correct age. Because of this feat the young campers held him in high esteem.

Charles was apparently the only one that suffered from having consumed portions of the sausage. He was sick for three days after the incident.

Wood Eye

GEORGE AND I WERE selected to run the evening recreation program for our school district. The students were expected to pay a small amount of money before they could be considered to be part of the program.

They were all allowed to listen to records, play basketball, table games and roller skate. We were expected to maintain order both inside and outside the school's boundaries.

The program ran Tuesday, Wednesday and Thursday evenings from 7:00 pm to 10:00 pm and lasted for 8 weeks.

I was asked if I would be interested in playing some 3 on 3 basketball in the gymnasium. I agreed to participate. I removed my eye glasses and placed them on the stage and told them that I was ready. We had not played for more than 10 minutes when retrieving a rebound, I attempted to dodge him but I was not successful.

He swiped at the ball but missed. He scratched my eye instead. It felt like I had just been hit with a large hammer. The pain was intense and continuous.

I notified George that I had caught a finger in my eye and asked him if he could handle the kids by himself. I was uncertain how I could make the trip home since my damaged eye kept closing and the good eye followed suit, leaving me sightless at times.

I was able to drive the car home and decided that the best treatment would be to go to sleep. This was not the ideal treatment because I remained awake more than I slept.

I realized I needed professional help. I called the doctor's office to tell him what had happened. He instructed me to come to his office as soon as possible.

When I arrived, he placed me in his chair. He told me that 90% of the outer eye membrane had been abruptly removed and hoped that he could save my eye. I told him I would do anything he suggested to save it. I had to wear a patch over the eye for a week. When I returned to see the doctor, one week later, he removed the patch and stated the eye had healed itself almost completely. The eye was still bloodshot but he assured me that this condition would disappear in time.

He noted that my eye's vision had improved during the week that it was covered. I did not tell him that my vision was blurred.

At work during my daytime job I told no one about my condition. I had to scan my test results by using a ultra–violet wand daily. This wand works best in a darkened room. I had to rely on my good eye completely to fulfill the necessary task.

I finally got full vision back towards the end of October.

Glazing

I ARRIVED ON CAMPUS three weeks before classes were to commence and decided to look for a job. I had experience installing glass on a greenhouse roof.

The job was dangerous and not easy to perform. The threat of making bodily contact with the glass was always present. One slight miscue could result in serious injury to the installer.

I started in the middle of the week and was given Hans as my helper. He was in good physical shape but had never installed glass on a greenhouse roof before.

His first step up the ladder involved a pair of wooden shoes. He was able to climb the ladder with relative ease provided his only concern was his body. Once he had to rely on his body to transport his tools and materials, he failed miserably. He was extremely clumsy. His second set of foot ware was a set of thongs. They were worse than his wooden shoes.

It was agreed upon that I would work solo. The only phase that I might require assistance would be to help move the greenhouse ladder from one bay to the next.

The ladder should be long enough to extend from the base of the ridge vent to the drain along the bottom of the cat's walk. For reasons of safety, the wooden cleats should be anchored with wood screws to the ladder itself. They are positioned on the top surface of the ladder.

The distance between the cleats should be more than the length of the glass. The supporting cleat at the ridge vent is attached to the underside of the ladder and is flush with the surface.

It is thickest, yet less in length than the other cleats. It is designed to slip into the area created by the open ridge vent.

The next three paragraphs might help those greenhouse owners that are intent on repairing and replacing the glass in an older personally owned greenhouse.

Polymer plastics have replaced the glass in new as well as old greenhouses. The cost of single panes of glass may prohibit their being used for replacement.

The advantages of plastic over glass are:

1. Less expensive
2. Panels cover a much larger area than a single pane of glass.
3. Panels weigh much less
4. Replacement is much faster
5. Tools needed for replacement are less
6. Personal safety is better–no glass cuts

Your county agent should be able to answer your greenhouse questions.

One thing to remember, with respect to glass to glass replacement, the higher pane of glass is lapped on top of the adjoining pane to assure water runoff.

Landfill

I WAS TOLD BY the girl campers that every night after dark strange noises were heard on the right side of their cabin. The sounds were not audible on any of the other three sides.

I decided to check out the noises after dark. The girls were correct in their observations. The strange sounds were audible on one side of the cabin but not on the other three sides.

I was certain that the sounds were made by bats. I also informed them that I would have to report my findings to the proper authorities who would close the camp for the balance of the season thus preventing the campers from having a great vacation.

I informed them of the consequences if this information became public. I told them to keep our secret and allow me some time to see what I could do to solve this problem.

I paid a visit to the camp's maintenance building to converse with the superintendant. I told him of my findings and asked him if he knew of any solution. I was aware that there was a pesticide designed specifically for the extermination of bats.

He told me that his shop did have this pesticide in aerosol spray cans. He offered these cans to me since he was in agreement with my plan of attack.

The siding pieces were installed from bottom to top horizontally so that each piece was tightly nailed along its bottom edge and made to lap loosely at its top edge thus providing a gap between adjacent siding members.

Through this narrow gap was enough space to allow access to the area behind the siding. This was where they slept during the daytime.

I told them that bats were nocturnally active at night and docile during the day. I also informed the campers that some bats were carriers of rabies, a disease more common in dogs.

The plan of attack was simple. We would spray one gap at a time and wait to note any activity beneath the siding. The scenario was the same for each animal extracted. The pesticide was fast acting and forced the legs of the animal to protrude first.

The gap was sprayed a second time forcing the bat to further extend its body out of its resting place. The bat's body would shake for a minute or two before it would go lifeless. It was then -passed to the younger campers who beat any remaining life from it.

This same procedure was repeated over and over again until the last bat had been exterminated. Their bodies were placed in thick plastic bags and then transported to a landfill.

The entire operation took less than three hours to complete. We disposed of 26 bats and were elated that none of their relatives returned to the site.

I realize that I took a big chance in executing my plan but can look with satisfaction at a job well done.

Guard

Most school districts close down for a week in February, usually attempting to honor one of the president's birthdays and also to anticipate when we would experience a heavy snowfall. My vacation was different from my kids so I decided to work as a guard.

My first assignment was to direct traffic at the main gate. I would direct drivers of each large flatbed delivery truck how to traverse the sloping roadway to the level storage area below, where the supplies should be deposited with ease.

My operation was going smooth until it started to snow. I was aware of what effect a small amount of snow would have on the delivery process. I had created a dilemma with only three trucks involved.

Within five minutes the head boss appeared at the main gate to relieve me of my duty. I was transferred to a side gate having no traffic what-so-ever. I remained at the no action gate until 12 noon when I was again transferred to a side gate having no traffic. I was stationed there for 2 hours before I was transferred to the carpenter's shed on the grounds of the reactor. I was given 3 distinct duties. The first was to observe and report any activities that were happening in and around the construction shed.

The main boss gave the directive but did not tell how or where to communicate this information.

The second duty was to inspect some large pipes exposed above the ground. They had been the object of controversy and the state police had been called to quell the uprising. Two separate unions were in conflict over ownership of these pipes.

The third duty was to receive a printed record of each time clock that had been strategically placed within the reactor. The entire operation would consume less than 5 minutes of time. The time scheduled was to be taken every hour on the hour.

Any suspicious activity was to be reported immediately. I never asked who or where was this information conveyed to. Fortunately this situation never came up.

I became quite friendly with an older man who was assigned to the carpenter's shed. His duty was to maintain cleanliness in and around the shed. He asked the construction superintendent to place him on the labor list for overtime for Saturday.

The super acted strange to this request and told "Captain" that he could not continue to put his name in for overtime. The "Captain" assured the super that this was okay with his super. Later, when we were alone, the captain asked me if I knew why everybody called him captain. He told me he and his broom would earn $18,000 for this year. He implied that this was payback for his former occupation and for keeping silent all these years.

Tennis

Main central high school had a reputation for having an excellent tennis team. They frequently would chalk up 20 to 25 victories from dual matches before they incurred a loss.

The teams would be willing to travel 60-65 miles to encounter a good opponent. They preferred home and away schedule with every team they played.

On away matches the team members followed the same scenario. They were excused from school early so that they would have ample time to complete the matches. The team, consisting of 5 members would be divided as follows: the best 3 members would play the #1, #2 and #3 singles. The remaining 2 members would play doubles against the home team. If the courts available numbered five, the visiting coach could decide whether to play five single matches or three singles and one double. This situation should be decided upon long before the day of the match.

It is understood that the home team is to supply one can of new balls for each match in progress. As soon as the pairings had been decided upon and the court determined for each match, the tennis could commence. I was totally surprised when the home team's coach asked me to give him a time when I should reappear.

I told him I was not leaving and felt obligated to stay with the team especially if one of the team suffered an injury. I asked the coach if this was customary behavior by my predecessor. He confirmed that it was and sometimes all of the boy's team waited as much as an hour for him to return.

On some occasions the smell of alcohol was easily detected. The thought of driving 60 miles with a car full of teenage boys while being somewhat impaired frightened me. This same scene repeated itself every time I had an away match.

Senior Day

SENIOR DAY WAS A long standing tradition at our high school. The school authorities decided long ago that they would set aside one day in the calendar year for the members of senior class. It had to be understood that those allowed to go on the trip must be in good standing with all the components of the school district.

Each senior was responsible for supplying the principal's office with a signed note stating that their son or daughter would be held accountable for any damage incurred. It had to be signed by both parents if they lived in the same home or by only one parent if only one parent was held legally responsible.

The time frame given for this note to be returned was any date two weeks prior to senior day. They received a complete listing of all the activities they could enroll in while at the resort. If the activity was involved with the expenditure of monies to maintain it, the cost was passed on to those seniors who chose to participate in that activity. For example, there would be a charge for horseback riding. The resort owners asked for a commitment and requested that these money items be taken care of prior to senior day.

I made a mistake by signing a contract with the suppliers of "city buses" to use their buses instead of the touring type bus.

It was an attempt at saving money and it back fired on me. Not only were they not designed for mountainous terrain, they were slow moving on level land. At times, one bus would take turns being the lead bus and then reverse the configuration to be the bus doing the pushing. Bus #1 would start the procedure as the lead bus being pushed until it reached the summit of a mountain peak.

The boys were informed that the resort owners had set aside a small building for the changing of clothes and anything else they felt like doing. I told them that I did not want them to destroy any item within the confines of the building nor the landscaping itself.

The girls building by design was some distance from the boy's. I told them that they were not to destroy any items, either inside or outside the building. I gave them my blessing and told them to enjoy themselves. I also told them that I would come to their defense if they required it.

I was called on the p.a. system for anything which the seniors felt was a problem requiring my assistance. Most of the problems were slight in nature and were solved in a heartbeat.

Then the finale of all problems came to a head with the discovery that someone, probably drunk, had taken a 9-iron to the walls inside the building assigned to the boys. It looked like a war zone.

The resort owners were there by the time I arrived. They wanted to know what suggestions I might have to repair the damage. I very quickly responded to their demands. I gave them two options. The first was for me personally, plus two of my buddies, to arrive at the resort the following day with the intent of making the walls whole again.

A second solution was to take the room deposit ($300.00) and use it to pay for the repairs. They immediately accepted my second solution with one concern, would this solution be overlooked by the school accountant. I convinced them that items like deposits are frequently overlooked. I told them I would keep in touch, by phone, to inform them of the outcome.

During the ride home, I discussed the situation with all the seniors and asked that all of them not mention what had happened at the resort. I was confident that none of the seniors would spread the word.

I was called the first thing Monday morning to the principal's office to discuss what happened at the resort. I told him that I felt that I had acted in a responsible manner and was confident that the matter had been resolved. He agreed with me.

Model

THE THOUGHT OF BUYING a model home is very appealing to certain home buyers. Most builders view the model home at the time of the sale as similar to a used car deal.

The buyer is expected to pay top dollar for every item that comes furnished with the house. For example, carpeting that was installed upon completion of the house's interior would be charged at the same price that the builder had to pay originally. What many people fail to realize or acknowledge is the number of years of wear and tear which the carpet had endured prior to their buying the model.

Spare parts are readily available from the models with the understanding that these parts would be replaced whenever they are needed.

Cabinets with a white-washed finish make their replacement difficult because of their exposure to the rays of the sun. Other stained finishes do not seem to present a large problem.

Berber carpeting may be difficult to clean and be appealing to the homeowner. One house where Berber was installed and subsequently cleaned, the carpeting displayed a mixed range of colors, some which may have resulted from a reaction between the carpet, the padding and the cleaning solution.

To retrofit the electrical circuits from model to residence usually requires some expertise, especially the master bedroom. While it exists as a model, there are additional outlets and switches installed mainly to keep the master bedroom more illuminated for display purposes.

In-ground swimming pools, over the years, may have developed cracks or leaks making it difficult to maintain a desired water level. Realize that the builder is probably aware of any problems but may not want to admit them to a potential buyer.

Most homes seldom present a problem with the plumbing. Model homes are different. The plumber is rushed by the builder to complete his work and thus some items are ignored i.e.; hot and cold lines reversed inside the home.

It should be understood that the builder wants the upper hand when dealing with a potential model buyer. Remember, a promise by a sales agent is not valid.

Stone house

IN REVOLUTIONARY DAYS, FARMERS picked up field stones that lay hopelessly in areas destined to be productive farmland. They consistently stacked these stones into walls. The stones also were used to designate boundary or property lines.

The walls were commonly 2 foot wide by 2 foot high. The height of the walls was dependent upon the abundance of rocks found in that location. The most preferred rock alignment was to stack two smaller stones on one larger one. The next course was to reverse this pattern, to stack one larger stone on two smaller stones.

They recognized the importance of not aligning all the joints in the same vertical plane. The same was true for the horizontal layer which may have been determined by the size of the available rocks.

The rocks may differ in color i.e. black, gray, red and yellow. They make an awesome display when blended together.

A woman called me and asked what I would charge to finish her home with fieldstone. Her husband had started the project but died before he completed the entire house.

She told me that she had an endless supply of fieldstones in a plot adjacent to her home. I arranged a time to meet her to determine if I was interested or not. Her husband had a mixer which she would not let us use. I could either rent a mixer or mix by hand.

I gave her a ballpark price to complete the job not knowing if I was too low or too high. She accepted my offer and stated that her husband would be delighted to have the job finished.

My younger son, age 12, was responsible for keeping my older son and I always supplied with fresh mortar. He employed three wheelbarrows to perform his mixing. I was not certain whether he could endure the job but he was ready to go each morning.

When we completed our task, the woman thanked us several times. She told us that she had contacted more than a dozen individuals and not one was interested. To me, it was a challenge.

Reactor

I TALKED WITH MY neighbor about my part-time job as a guard at the nuclear power plant. I asked him if he was ever involved in the construction of the reactor in any capacity.

He asked that I not convey any information about his answer to anyone. Curly told me that he and his friend Riley, both union carpenters, had full-time employment since the inception of the project. These two had the distinction of being on the payroll from the very beginning but never drove a nail.

One of them would go to the reactor site on Monday mornings, sign in for both of them, while the other man would be present on Friday afternoons to sign the completed timecards and leave the scene with their pay.

They were in fact building a "spec house" for a private investor. Not a bad deal for two country boys.

Because of the shortage of man-power, I was asked to perform two separate tasks. I was asked to punch the clocks throughout the reactor as close to the hour as possible. This function normally would consume 5 minutes of the hour.

My second task was to leave the reactor area to check on a barge that was anchored to a mooring on the bank of the river. I was expected to climb up on the side of the barge and peruse the deck for any activity.

Once this task was completed, I was to punch the clock located within the tiny construction trailer nearby. This trailer was equipped with a space heater designed to keep the trailer warm. Even though the day was warm for February, I only entered the trailer to punch the clock.

At 3pm, the second in command came by to inform me that I could work a double shift. I thanked him but told him I had had enough for one day.

I was lucky my replacement arrived before 4 pm. I explained what was expected of him and said goodbye and good luck.

One of my buddies that worked full-time informed me that my replacement had been overcome with carbon monoxide fumes. They found the man at 8 pm. Thank you, Lord.

Christmas decorations

SEVERAL STUDENTS WERE CURIOUS about how I planned to spend my vacation. Most of them knew that I would be selling plants and flower arrangements at the florist shop. My work schedule included two or three days beyond Christmas.

I told them that I would be decorating the tree Christmas Eve after I finished selling at the florist shop. They asked if they could come to my house to aid with the decorations. I told them yes, never expecting them to follow through.

I arrived home about 6 p.m. and placed the tree in its stand. We explained to our son that we were helping Santa with his work by erecting the tree and placing the lights throughout the tree.

I was quite tired after hanging all the lights so I decided to rest before I continued with the balls and garland.

The phone rang and it was five of my students. They needed directions to my house but first they wanted to make certain that I did not mind them coming to my house.

I assured them that it was fine for them to pay us a visit. I was very glad for help. They confided with both my wife and I that they had never decorated a tree. They stated that the closest object they had in the Jewish faith was the menorah or festival of lights.

The lighting of eight lights, one per day, was a happy occasion. Its tradition suggested that the youth in the household received one gift per day for eight consecutive days.

My wife started the girls off on their decorative junket by handing each girl a Christmas ball to hang on the tree. She told all of them they would need a metal hanger as the means of hanging the balls on the tree.

Not one of them connected with what my wife was telling them to do because all five balls were on the floor rather than on the tree.

She gave a metal hanger to each girl, plus instructions on how to use it. Their excitement and enthusiasm were pleasurable to see. All of them commented on how much fun they were having.

Once they completed hanging all the Christmas balls, they were ready to place the garland. Most of the garland placed on the tree could be either silver or gold but not both.

All the girls thoroughly enjoyed themselves and expressed their desire to return the next year for a command performance. I told them that their efforts were greatly appreciated and would welcome them again with open arms.

Roof repair

RARELY IS THERE A complaint about a ceramic tiled roof. Their completion involves a series of steps. The roof trusses are nailed in place, and then the plywood roof sheathing is nailed to the trusses. The heavy roof paper (90 weight) is then applied across the sheathing.

Each piece of sheathing must be large enough to span across three trusses. Next, the roof tiles are assembled on top of the roof paper. The tiles are basically cosmetic and are not responsible for water runoff. That is the role of the heavy paper.

The builder received from the building inspector's office a letter requesting a step ladder be positioned near the attic access to allow the inspector easy access to the attic and furnace.

I made certain that I abided by the request by placing the step ladder in position. I placed the ladder in close proximity to the access cover and noticed a series of light rays once the cover was lifted. They were being emitted around the perimeter of a small section of roof sheathing. I immediately notified my supervisor of the problem and requested that he check it out thoroughly.

He checked the roof from the outside to see if there were any apparent differences in the arrangement of the tiles. He then proceeded inside, entered the attic and realized that my observations were confirmed.

The tile installers had apparently been aware of the situation and felt they could cover up their mistake. A 4x4 piece of sheathing was missing but they managed to assemble all the other necessary components in their correct alignment giving the appearance that all was okay.

My boss called the roofing company to explain our dilemma and strongly suggested that this mistake be corrected immediately. He then called the building inspector's office to postpone a series of final inspections.

It is scary to imagine what would have taken place if this problem was not identified before the house was occupied. The roofers may never have encountered another situation of this magnitude.

Shell casings

I SPENT ONE OF my summer vacations working in a drop forging plant. While there, my mission was to cut solid brass cylinders measuring 6" in length by 2 5/8" in diameter. Each eight twenty foot solid brass cylinders were tied together four in a row and two high.

To insure a uniform cut, the cylinders had to be cut slowly to remain consistent in length. Cutting them rapidly caused all kinds of irregularities.

This was a rush order from the government. The factory personnel were anxious to comply with their wishes. Because of their demands, the operation was ongoing 24 hours per day.

To insure a uniform level cut, 8 brass pieces, twenty feet long, were tied together and lifted with several chain hoists equally on the uncut brass. The chain hoists were periodically moved every foot to achieve uniformity.

Those brass cylinders cut to proper size were placed in a reinforced cardboard box ready for shipment.

I made certain that the cylinders I cut measured the desired length. I frequently used my tape measure to check the length of each piece.

I was called to the office to explain why the cylinders ranged in length from 3" to some that measured almost 8". I was annoyed that my ability was questioned. I reminded my superior that I was not the only one cutting the brass. The employees from the other two shifts may be those causing the variation in length. I also informed him that whenever there were several people performing the same task the overall outcome cannot be guaranteed.

It is my belief that the other employees were not as caring and meticulous as I was.

To identify the shift (employee) responsible for the variation in length, have the 2nd shift perform alone the first evening. Next, have the grave yard employee cut the brass. This procedure would not only identify the guilty party but also exonerate the one doing the job correctly.

I changed from cutting brass to helping another employee fulfill his task commonly called "cold strike". This technique took each small steel piece, placed it between the two dies, and restruck it in its cold state without the need for heat. It was small piece of steel that had previously been heated and drop forged. This component, found in motor vehicles, was first cut to an appropriate size to minimize waste and then heated in a furnace. It was then placed between the two dies, one as the stationary die or the bottom die, the other as the top or moveable die.

The operator of the drop forge machine had to trip the pedal with his foot to cause the top die to proceed downward striking the stationary die. The cold strike was done to correct a minor flaw in the forged steel.

This procedure was paid on a "piece work" arrangement so the more pieces we corrected the more money we would receive in our paychecks.

The machine was to be blocked between the dies whenever it was being inspected or cleaned. Correct procedure requires a 4x4 block placed on top of the stationary die and a 2x4 post between the two dies.

I discovered after I left and returned to school that my friend, the operator, did not block his machine. He lost three fingers because he was in a hurry to continue with the cold strike procedure at the expense of his own personal safety, proving that haste makes waste.

Hide and Seek

THE THIRD WAVE OF campers was to arrive in early afternoon, in time for dinner at 1 p.m. I made an accounting of the staff to see whether or not they were ready to deal with the new campers.

Each camper should know which bed is his or hers. They must accept the bed's location and not change it for any reason. This is important whenever one of the campers turns up missing. This became reality the very first night these campers were sent to bed. A bed check was called for the two cabins housing boys.

It was soon determined that one of the boys was missing. He was also the shortest of all the campers. He left his cabin and decided to hide in the woods surrounding the cabin on all four sides.

I decided to organize a search party. One of the items that each camper was told to bring to camp was a flashlight.

My wife, the assistant director, was to remain with the girls in their cabin thus freeing up the two girl counselors to help in the search of the missing boy. The other boy counselors remained with the boy campers.

One of the counselors suggested that our search party be enlarged to include most of the boy campers. I denied this suggestion and told him that I would feel more comfortable with one camper missing rather than a larger number.

We found the missing camper within 15 minutes. He was cold, frightened and scratched because of encounters with the underbrush.

I let him call his parents to assure them that he was alright but no one answered.

The very next night was a repeat performance. The only difference was his tree selection. I called the father and asked that he come to camp to retrieve his son.

Patio covers

AFTER SELECTION OF THE home and deciding what style best suits them, the next decision confronting the home owners is what type of patio cover they prefer most. In a community where new home building is thriving, the number of individuals willing to build you a patio cover is legion.

Some are both talented and knowledgeable while most seeking your money are not.

Many do not know that the attachment to the house should be wood on wood. The exterior wall studs must be exposed so that the house plate is nailed to the studs. It is best to expose the exterior wall surface 4-6" above the top of the house plate so that it will accept the metal flashing. If done properly, the flashing will direct any water running down the wall to reach the flashing and be deposited on the patio roof and not enter the house interior.

Whether you decide to build your own patio cover or hire a licensed contractor, the first order of business is to file for a building permit. Too many home owners and non-licensed contractors ignored this very important item. Later, when the house is up for sale, they are required by the loan officer at the bank to produce this document.

Before applying for a permit, it is advisable to decide all the features that you feel are important to make your patio cover one that you will be proud of.

Rigging Boats

I WAS HIRED BY a local boat shop for some part-time employment. The owner had given me a job when nothing else was available once before and I felt indebted to help him out.

My part-time plate was quite full but I assured him I would make some time for him. I was already tutoring students, coaching tennis, working as a salesman in a discount store in the garden department and writing an occasional magazine article. My last job of the evening was in sales. The store closed at 9 pm all evenings except Saturday when it closed at 10 pm.

I told him that I could be at his shop by 9:30 pm and be ready to work. My job was to install windshields and canvas covers on new boats that were scheduled to be ready for delivery the very next day.

One night when I arrived, he apologized for having promised four boats for the next day. I told him not to worry. I could count on four hours sleep if I made it home by 2 am. I just asked him not to make this a common practice. He liked to barbeque after I arrived and would gladly share his food. That scenario probably kept me going to the wee hours of the morning.

I completed the four boats with time to spare. I made certain that the shop was locked completely and then proceeded home.

I found out the next night that one of the four boats I worked on had been stolen from the boat yard. It had been placed outside the boatyard to gain additional space needed to work on the other boats. It was positioned on a trailer. The thief simply backed up to the trailer connected it to his vehicle and drove away.

Lawn mowing

ANYONE WHO DOES LAWN maintenance for a living has my blessing. Mowing the same lawn week after week has to be monotonous to say the least. I worked for two summers for the same landscaper. He believed that the various tasks associated with the landscape should be kept simple and whenever possible be varied to make each segment appear different from the week before.

He would mow the front lawn from north to south (parallel to the street) and the next time the lawn was mowed perpendicular to the street. A third method of mowing was called "cross mowing". This practice helped smooth out the rough spots giving the lawn a well manicured appearance. Cross mowing involved cutting the lawn on a diagonal (a 45 degree cut).

To ensure a well managed lawn, it's strongly recommended that the lawn surface be swept with a lawn sweeper and that the clippings be collected and discarded.

The two employees sent to do the mowing were promised by the foreman to have additional help to complete the job before they quit for the day. Fred and Carlos, one mowing, the other sweeping, had nearly finished the front lawn and were moving the owner's equipment to the backyard when I arrived.

Since I was the new person on the job I was given the company's sweeper to finish the front lawn. I was not overjoyed to receive this assignment but accepted it none the less.

I became aware of noises and shouting from the backyard. Soon after, the housewife bolted from the front door, ran up to me, and stated that the shorter of the two had a twelve inch knife that he was attacking the larger man with. She wanted me to intervene and stop the fight before somebody was cut. I told her to call the state police. They arrived at the site within ten minutes. The two individuals were then handcuffed separately, placed into two vehicles and were transported to the barracks.

They remained in the custody of the police for three days. During those three days each of them lost their job and told that neither of them would warrant a recommendation.

I told the housewife that I would return the following day to complete the mowing of the lawn. Upon entering the garage the next morning, I noticed some disturbing items in clear view. There were loose coins on the garage floor and on the base of the foundation plate. In another location, dollar bills were placed between the 2x4 wall studs. Bait? Yes, but not enough to go to jail for.

One of her neighbors was the individual who called her about the knife fight in her backyard. The neighbor had spent some length of time behind her curtains before she called to report the fight.

During normal daytime activities, the knife was probably encased in a leather sheath to protect Carlos against injury. The police did question him about the knife but I was beyond the ideal audible range when he answered the questions.

Another problem that landscapers face is urination. It creates a problem that most home-owners do not want to deal with. Some have a utility bathroom just inside the garage and in close proximity to the outside. I asked the woman if she and her husband ever considered allowing the help to use their bathroom. She replied no but could understand "my point of view'.

If the homeowners had a standardized permission form that they could have each laborer fill out completely, this procedure might ease the attitude of all involved.

Plumbing problems

MOST PLUMBING FIXTURES ARE installed having the cold water supply line on the right side and the warm water line on the left side. Some shower valves are equipped to turn 180 degrees out of position to correct this condition without dismantling the wall.

Sinks and vanities have separate cold and hot water lines which can be reversed beneath the cabinet by changing the shut-off valves. The same holds true for the washing machine with the extra requirement of a drainpipe being installed.

To prevent the possibility of "water hammer", the installation of an air chamber is suggested. It should be at least 18 inches long and be connected above the supply line. It is designed to minimize the noise created when a fast flowing faucet is shut down.

Sometimes in a newly constructed residence, the showers do not spray water as designed. It may have been caused by building debris that plugged the water line to the showerhead. The showerhead should be removed and the water turned on to expel the debris from the shower head. Prior to assembling the head to the water line, examine the head to determine whether or not it has been cleared.

Contracts

WHENEVER YOU EMBARK ON a project costing several hundreds of dollars or more, you should have a contract between you and the craftsman you decide to perform the services you desire.

The contract should spell out what you expect to be accomplished, paying strict adherence to the building codes in your area.

If a building permit is required, determine who files the necessary paperwork and how many inspections are required.

If you are contemplating being your own boss, and hiring day laborers off the street corner to perform, you will probably get taken. The mystic that most of these individuals are more talented is a myth. They are only more opportunistic.

Projects involving a building permit are invalid if the work was accomplished without a permit. Banks will reject mortgages whenever they can prove that the work was done without one. For example: a homeowner wants to replace his rotting 12 x 12 wooden deck with a larger 12 x 16 deck. He decides that he will build the new one without filing for a building permit. If he attempts to sell his home with the new deck, the potential buyer can question whether a permit was obtained and this could queer the sale.

A contract should help the client and the craftsman with a clear understanding of the work to be done. If the client forgets to inform the craftsman of his wishes, the information is never entered into the contract.

I built a stone wall for a client who never expressed a desire to

contained all the necessary components to make concrete. The truck held sand, gravel, cement and water in the proper amounts.

The owner of the truck asked me if I would pay the towing charge when the truck got stuck in the wet soil. I decided right there that we would have to use wheel barrows to fill the trenches. This method would take more time but it would be safer. The truck owner told me not to worry. He was slow this time of year and would not charge me for the extra time.

Even though we were forced to wheel all the concrete and deposit it in the soil dug trenches, the procedure did not consume as much time as I had anticipated. We poured two of the four walls and the two corners that connected those same walls in our first day.

We anticipated a problem with the one corner. It was necessary to construct a step down form not only to accept the concrete in the first pour but also to guarantee that the concrete in the second pour would blend and mesh with the first pour.

I was very grateful to have hired this particular concrete man. The first day he only charged me for the concrete we used and did not charge me for the extra time. The same was true the following day.

I drove by the house about two months later and saw that the addition was complete and it looked great.

Floor Pops

MOST CARPENTERS KNOW THAT they must thoroughly anchor the floor sheathing to the underlying floor joist or floor trusses before the carpet and its padding can be installed. Usually most framers reserve this procedure to a trusted employee who guarantees that the job is done.

The desire for speed of assembly can affect the final outcome of the flooring causing the characteristic "floor pops". If the sheathing is not nailed down immediately after it has been cut to size, you run the risk of ignoring the snug fit completely.

If this condition is not taken care of when the sheathing is installed, it will not be addressed by the carpet installer.

The solution to this dilemma is to have the carpet installer check the nailing patter of the sheathing. If it is not nailed properly, he should be expected to correct the situation.

Whenever the carpet has been installed and the furniture placed in the room, usually a bedroom upstairs, it may not be apparent that a floor pop will present itself right away. It may be several months before it is detected. The only solution to the problem is to remove all the furniture, carpet and its padding, thus exposing the floor sheathing. Then, and only then, can the job be done properly. Trying to do half of the room and then the other half proves many times to be less than perfect.

If done this way, a ridge usually forms in the center of the room.

It may either affect the padding or the carpet. In either case, one of them has to be trimmed slightly to make it fit properly.

"Floor pops" can appear in the vicinity of the top step of the stairway. This area has to be "stick-built" since the manufactured trusses may not fit the opening.

The task of framing the area at the top of the stairs should be given to an experienced carpenter who knows and understands how to frame this area. Over building is preferred to skimming on materials.

Sometimes there is a large void between two or more trusses that emits several different sounds when stepped upon from above. They may vary in intensity and pitch.

Traveling man

WE WERE ALL GATHERED outside the mess hall for our Sunday meal when my wife asked me if I knew where our 3 year old son was. I told her that I did not know his whereabouts but would check inside our cabin first.

She started to check before I did because it was my job to alert our counselors that our son was missing and to inform us if any of the campers were aware of his location.

My wife found him in our cabin. He had consumed 40 baby aspirin (the equivalent of 10 adult aspirin) and was in the process of devouring a 6-pack of Ex-Lax.

She let out a scream for me to hurry to our cabin. She showed me what he had ingested. She explained the location of the items hiding place. I was not aware of the hiding place but apparently our son was aware of where the cache was located.

The two nurses associated with the camp were there for the meal. One of them told me that he would need to have his stomach pumped. The other nurse alerted the state police, telling them what route the Plymouth station wagon would be traveling and our destination.

The one nurse told me to keep my son awake and not allow him to fall asleep. I sat in the front seat holding my son in my lap. The head counselor would drive the wagon.

Our one way trip was 30 miles and it was difficult to keep him awake. I was able to keep him awake by pointing out objects in the landscape, some real, some fictitious.

Once at the hospital, the staff was quick to respond and wheeled my son inside to the emergency room. They placed him face up on a table and began wrapping him with sheet materials.

This procedure caused him to become fully awake and have him use his feet to kick the hospital staff. They assumed that he was more asleep than awake. They misjudged his strength.

Once the hospital staff released him, I was ready to return to camp. My wife had come to the hospital in another vehicle. She had fainted at the camp before she got into the wagon and we were told by both nurses to go ahead and they would care for her.

As soon as he saw his mother, he asked if she was all right. All of the campers were glad he was fine.

After our Sunday night meal, it was customary to gather around a campfire and to sing songs.

We were just getting organized when one of the campers came to me with tears in his eyes. He told me that one of the counselors from last year, who was in attendance, was attempting to pit two campers in a confrontation. He said that the other camper did not want to fight either.

I immediately approached the former counselor telling him that I did not want any of my campers to be intimidated. I told him to gather up his belongings since I was taking him home. Word spread quickly what I had said. They were so happy about my decision that they cheered.

The trip to his home was somewhat slower than my first trip that day. Ironically this trip was almost identical since the antagonist lived directly across from the hospital.

When we arrived at his house his family was seated on the front porch. His father asked if his son was injured or sick. My only comment to the father was that his son would explain what had transpired at camp. I told all of them that their son was welcomed at camp only in their presence.

Attic addition

I RECEIVED A PHONE call one evening from a home owner asking if I did attic additions. I told him that it is sometimes difficult to give an answer over the phone without seeing his situation.

I made an appointment with him for the next Saturday morning. He informed me of his wishes. He and his wife had 3 boys, 2 of them teenagers, the third was only eight. All three were occupying the same cramped bedroom and it was becoming more hectic with each passing day.

If he had the attic area converted to two bedrooms that would have solved many of their problems. He also did not want his real estate taxes to increase because of the addition. I informed him that he should apply for a building permit, but could possibly avoid applying for one provided no one in the neighborhood was aware of what was happening in his house.

I examined the stairwell and told him that the existing set of stairs had to be taken down and reinstalled so that the top of the stairs becomes the base of the new positioned stairs. The stairs would be turned 180 degrees.

This had to be accomplished before anything else was done. I told the owner that he was lucky because the existing stairwell appeared to be constructed with this thought in mind.

After the set of stairs had been repositioned to accept foot traffic from the center of the home they were braced and anchored in place. They fit the opening so snugly that no shims were necessary.

I told the homeowner that I would have to order a custom made set of stairs for his house. I had to measure the height of the opening (the rise) and the length of the horizontal stairwell opening (the run). They promised delivery the following Thursday.

I gave the lumberyard my measurements. When I arrived on Friday I was upset that my measurements were either ignored or they had in their possession a set of stairs with similar measurements they felt they could palm off on me.

I had to devote 2 hours of my time to customize the stairs to fit the opening. The stairs were 2 steps longer than they were supposed to be. My first step was to remove the two extra steps.

My next dilemma was that the width of the stairs was somewhat wider than the stairwell. I attached a length of rope to the riser portion of the top step and the other end was looped over a collar beam that was aligned some distance away from the top of the stair opening.

One person was to pull on the rope slung over the collar beam while another person guided the bottom of the steps so that the stairs were snug in the opening. With back and forth motion, the weight of the stairs forced the stairs into the snug stairwell, eventually achieving the desired result.

The steps were nailed in place on the closed wall side of the opening and braced beneath on the open side with two 4x4 posts.

I asked the homeowner if he planned to heat the new bedrooms. He stated yes as long as the cost would not be too high. My friend Tim would make an in-depth list of the materials needed to enlarge the heating system to accommodate the new bedrooms.

The only real difficulty was drilling the holes in the flooring to accept the heating pipes. We drilled all the holes in existing closets to hide them from view. I made certain that there were shut off valves to each bedroom.

For safety's sake, we installed a handrail on both sides of the steps.

The floor of the attic was covered with ¾" plywood that was nailed down with snug fitting joints.

We again had to consult with the entire family to see what each family member envisioned the floor plan to be.

The following had to be considered as necessary items to be incorporated into the floor plan:

1. A railing had to be installed around the top portion of the stairwell. The balance of the stairwell should be covered with ¾″ plywood to provide for extra storage.
2. Knee walls should be installed to separate living space from the dead space. The walls should employ 2x4 structure for the frame and be covered with drywall.
3. Additional collar beams may be needed to properly frame the ceiling area before it is covered with drywall.
4. The homeowner wanted to try his hand at completing the framing and installing the drywall. That request was all right with me.

Prefab House

PREFABRICATED HOMES OR MODULAR homes are preferred by some home buyers for a variety of reasons.

Most components of this type of house have first been manufactured in standardized sections which are ready to be assembled quickly in a factory. The materials are less expensive because most of the costs related to construction have already been dealt with.

My experience dealing with modular homes have led me to be cognizant of certain facts related to the delivery of the modular home to the delivery site before the home can be accepted by the purchaser.

Prefabricated homes require a crane to assemble the various sections. The components or sections are limited to a width of 14 feet to insure that they can easily be transported on the highway.

Before the home can be delivered to the site, a foundation must be constructed. It must be level. Its dimensions must be the same for both opposite sides of the foundation. If one side is shorter than its counterpart, the building may not be centered on the foundation.

If the distance from the outer edge of the longer side of the foundation to its complimentary outer edge is not as long, the house may under hang its foundation.

The first modular that I was involved with consisted of two halves which when joined together produced one small house. The first section meshed perfectly with the foundation. I made the assumption that the second section would fit as nicely as the first.

I was wrong. The crane operator lost control of the section and dropped it in the street. It was witnessed by about 25 neighbors. It did not mesh with the first section.

Whether the bump in the road was instrumental in causing the poor fit no one was quite certain. Some gentle persuasion and the proper positioning of wooden shims, the sections meshed perfectly.

Another feature that prefab owners like is that all the plumbing and the electric has been installed. The connection for both of the utilities is made at the building site usually close to the entry door.

Boat insurance

SEVERAL OF THE BOAT shop clients were frustrated with their boat insurance coverage.

The owner of a brand new boat had a practice of sitting on top of the driver's seat rear cushion and with his feet planted on top of the driver's seat. In his mind, this seating arrangement provided greater challenges then the conventional steering of the boat.

He was alone, as he frequently was, when he released the boat from its trailer and tied the boat to the mooring dock. He anticipated a fun filled afternoon with not a care in the world. He drove his boat parallel to the coastline to check the boat's ability to negotiate turbulent water. The boat handled every test its owner tried on it.

Next, he tried circles with his boat. Poor choice. No sooner had he commenced with the circles when his boat was hit broadside by a much larger wave. It caused the direction of the boat to change its driver intended direction and began steering towards the driver. The boat made two passes at the owner which were extremely close to making contact.

The boat's driver was relieved that the boat motor's propeller and the boat itself had missed him.

The next thing the run–a–way boat did was to head directly for the rocky shoreline as if catapulted there. The owner of the boat gathered any papers relating to this boat's identity. He reported what had happened to the boat salesman that had sold him the boat. The owner was alarmed when told that if he was not in the boat at the time of the crash the insurance company will not honor the claim.

Another boat owner had decided to make necessary mechanical corrections himself rather than having the repairs done professionally. Many boat owners make the attempt at repairing their own boat. They would be able to make the needed repairs at a substantial savings provided there was no incident resulting in additional damage to the boat. This happened to one boat owner who was attempting to simultaneously make two different repairs at the same time to his boat.

The one involved fiberglass repair to the boat's exterior, the other the necessary fuel line repairs required to free a clogged fuel line. He apparently got confused with the two organic solvents he was using for the two repairs.

He was surprised when one of the solvents caught on fire. He was not prepared to combat a fire having no fire extinguisher at the repair site. He also was denied his insurance claim.

Highs versus lows

I SPENT ONE SUMMER working as a laboratory research assistant in a chemical establishment. My assignment was to test new pharmaceuticals to determine how well they were absorbed in the small intestine.

Two rats were used each day. They were first anesthetized prior to surgery. One incision was made at both ends of the small intestine, one at the start the other at the end. Plastic tubing was inserted into the small intestine at both ends and tied with nylon thread.

The amount absorbed was determined by subtracting the amount received at the end of the intestine from the amount entering the small intestine. The rats were placed on a board to prevent them from moving.

The initial period of perfecting the technique was frustrating. For acceptable data, the technique had to be performed on two successive days. Quite often, one of the rats would die before completion of the test or a blockage might occur in the intestine or the fluid being tested would swell up inside the animal's body.

I expressed my frustration with the testing process and lack of positive results. My boss told me not to worry and to be patient. He explained that I had to be cognizant of my biological highs and lows. Some days I operated with perfection while other days my talents were at the other end of the spectrum.

Most individuals are aware of their abilities to perform admiringly in today's society but do not associate it with their highs and lows.

He gave me an example of a career chemist with his daily schedule can work for a lifetime (40 years) and have only one chemical hit. Some do not have any at all.

He also acknowledged that the attainment of a marketable pharmaceutical requires the services of many supporting cast members before it can be sold in the market place.

He gave another example of highs and lows. Commercial airplanes have designed a system to have the pilot fly the plane during his highs and act as co-pilot during his lows.

Home Addition

I RECEIVED A PHONE call from a homeowner who wanted to learn the feasibility of having two bedrooms constructed over his garage and the price to complete such a project.

I told him that I would meet with him at his convenience. The present house had 4 bedrooms and the addition would result in 6 bedrooms, all on one level.

I told him that the roof rafters presently used for the single garage roof would be reused as the rafters covering the new addition. This suggestion reduced the expenditure for the project. Likewise, only two windows will be required since the two existing side windows will be repositioned on the outer new walls, 1 in each bedroom.

I informed the home owner that no matter who he chooses to do the construction, he must submit a layout of the addition to the building inspector's office, showing 2 hot air registers, electrical outlets, light switches and closet locations.

I suggested that the entry door to each bedroom be placed on the common wall between the two rooms. The walk-in closets would occupy most of the remaining space along this common wall.

I gave the homeowner a ballpark figure of 10-12 thousand for time and materials for the entire project. I explained that I might be off in my estimate and it would involve some measurements to be more correct.

I explained the estimates for the electrician, the air duct installer, the insulation and the dry wall would all require more thorough calculations. I would telephone each jobber to ascertain the correct amount for each service.

To do the construction, the following stages would be performed:

1. Install floor joist
2. Apply floor sheathing to the floor joist
3. Dismantle the existing roof i.e. shingles, sheathing and rafters – dispose of the roof shingles
4. Use 2x4's to frame the three outside walls, and 2x8's as the window headers
5. Nail 2x6 collar beams to the top double 2x4 plate
6. Frame the gable end with 2x4s extending the roof using lock out rafters
7. Employ a 2x6 as the ridge beam for the roof
8. Nail the rafters to the top plate and to the ridge beam
9. Apply the fascia to the cut end of each rafter
10. Apply the roof sheathing to the rafters
11. Apply the drip edge and then cover the roof with 15 lb. paper
12. Install new roof shingles to the sheathing

Once the roof was completed, the wall sheathing can be nailed to the 2x4 exterior walls. The four windows can be installed as well.

Entrance to the new bedrooms must be accomplished prior to the installation of all the wall sheathing. This is done so that there is access to the new bedrooms from the existing hallway.

Begin by removing all the drywall at the end of the hall. This procedure will help define the size of the opening and dictate what materials need to be removed and what materials will be required to frame the opening.

Start this phase by removing all the exterior barn shakes from the old bedroom walls. Follow this by removing the wall sheathing on the same wall. This material can be reused on the gable end of the new addition.

Insulation can be left intact on the existing wall or repositioned on the new side wall.

Frame out the hallway extension to accept the drywall, leading to both bedrooms. Measure the amount of insulation needed to complete the wall areas plus the bedroom ceiling areas. It is possibly more cost effective to hire an insulation contractor to install the insulation.

The electrician selected can run the rough electrical service throughout the newly formed 2x4 walls at this point in time. Likewise, the heating contractor can run his duct work from a source in the garage.

To seal the exterior, 15 lb. black felt is stapled to the wall sheathing. Most of the sheathing removed from the former gable end can be repositioned on the new gable end.

Barn stakes are used to cover the felt over the entire exterior surface following the same alignment already employed on the house.

The closets for both bedrooms can be framed and positioned using the common wall as the rear wall of each closet. Sliding closet doors would minimize the available space.

The interior walls can be covered with drywall or paneling. The ceiling must be covered. The joint lines must be covered with tape and joint compound initially before additional coats of compound are added. Each application of compound must be sanded smooth before adding another layer.

The walls receive the same treatment as the ceilings and can be painted once the walls are smooth.

The installation of trim molding makes the project complete. It consists of base molding surrounding the entire floor, door jamb molding instead to trim out the two bedrooms and 4 windows and sliding closet doors.

Sod vs. seeding

BEFORE THE ADVENT OF commercial sod farms, seeding a lawn was the only way to achieve success. Now there are methods and techniques employed to produce sod at an alarming rate.

Sod is produced in the spring and again in the fall. Fall is the most favorable time of year to seed or sod a new lawn. Either one can be done in the spring season but germination of lawn seeds is hampered by the competition with weed seeds.

The owner of the landscaping business had been approached by a local insurance executive to sow his property with a mixture of Kentucky bluegrass and fescues. The entire property was in excess of three acres.

The landscaper told the architect that he wanted plant beds surrounding the perimeter of the property with the balance of the property devoted to a beautiful lawn.

It was acknowledged by all parties concerned that the rear yard would be the starting point. The front yard would be the last area to be sown.

If the property was to be watered by an automatic sprinkler system, the first order of business would be to designate the location of each sprinkler in the entire area. They should be placed on a large map somewhere within the confines of the garage.

Sprinklers would service the massive lawn areas while drip irrigation would be relied upon to water the plants. Location of all the sprinklers should be noted on the large map. One or two of the drip or bubble irrigators should be sufficient to water each side of the balled plant.

It is important to check all the water lines involved with the lawn to ascertain coverage as well as quantity of water to be delivered. This has to be accomplished before the seed bed has been leveled and has received any additional soil and fertilizer.

At this juncture both the sodded lawn and the seeded lawn have both undergone similar treatments.

Sod is grown on a concrete platform. Prior to receiving the soil and its additives, the welded wire used to cut the sod must be examined to check its operation.

The platform receives all the components; the rock free soil, the organic matter the lime and the fertilizer to a depth of 3 inches. A concrete ready mix truck is best employed to blend all the components before they are added to the platform.

Sod is usually purchased in rolls 18" wide by 6' long. Each roll is designed to cover 9 square feet of lawn surface. Each component is rolled on the ground and great care must be exercised to insure that all joints are installed with no space between them.

Sod has the advantage over seed for the following reasons:

1. It is very fast to install
2. It is usually weed free
3. Its growth is not delayed by a late sowing date
4. It is great covering hard to seed area

The entire rear yard had been completed and mowed at least twice before the front yard was started. The intention was to continue to sow the lawn areas with seed.

Time and materials both ran short as the overall project was approaching completion. Another day was necessary to finish the seeding in an area directly outside the front door.

The landscaping foreman assured the home owner that the finishing touches would be accomplished the very next day.

Instead of scheduling the work in the morning, the landscape owner decided to have his entire crew arrive at 3 p.m. Some of the employees began spreading the soil in the non-sown area. Other employees were concerned with the abundance of weeds and removed the visible ones. They were told to ignore them and were asked to help grade the area so that the landscape owner could be paid in full.

The homeowner held back payment on a wait-and-see approach. We were forced to return to the area and reseed it.

Tutor

SOME OF THE STUDENTS have the ability to concentrate on the classroom activities and absorb a major portion of the subject matter presented to them. Others are not and this requires additional help from other students, from their teacher(s) during after school sessions, or from very sincere and gifted tutors.

They may require tutoring once a week to just stay current. Others need a tutoring session just before an exam.

The normal tutoring session takes about one hour. Both the tutor and the student are expected to be on time, to have the necessary materials, i.e. textbook and subject notebook, and to have a quiet study area with no distractions.

The parent(s) always wanted to know why their son or daughter exhibited classic and chronic signs of poor study habits. I quite frequently discussed this problem and offered them solutions.

I have encountered the following:

1. Student more concerned with the actions of his pet boa constrictor than being tutored.
2. Mother and son argument about how and when he should study.
3. A bull mastiff was present and disruptive during the entire session.
4. Pair of Siamese cats either chasing each other or climbing their clawing pole.

5. Attempted to tutor 6 girls at the same time-number dropped to 2 after 3 weeks.
6. Had to enjoy a snack either before the session began or ate during the complete session.
7. Offered home study to students that had been out of school because of an illness-one father wanted to have me join the John Birch Society and spent 15-20 minutes each session attempting to persuade me to join.
8. One of the finest cross country runners who suffered from an inferiority complex among his peers.
9. A motor-cross enthusiast who continued to tell me about how well he performed.

Two of the nine students were unable to cope with their existence and decided to end their lives. One with a gun, the other chose a busy parkway.

To prepare students to take the state exam, I played host for an elaborate study session. It involved a swim in the pool, picnic on the patio, followed by a thorough question and answer session.

Several of my students were destined to fail without intensive review. I was happy to have many of them in attendance and have them pass the exam, given 2 days after our review session.

During my last year of teaching, I was informed that because of decreasing enrollment, I would be expected to teach a subject that I was certified in, but had never taught. I discovered that the adjoining school district was offering an eight week course and was looking for a teacher. I applied and got the job.

Choosing the job enabled me to help my present class of twelve students, and gave me the time to organize the various topic areas into easily understood information.

All had taken the full year course and failed. Three of my students passed the exam.

I had a student in my class that came to my desk after his classmates had left the room, with tears in his eyes. He told me that he was an army veteran who had a shell explode very close to him. The end result was that his hearing was affected. He said that he was proficient at reading lips, but not if the teacher chewed gum. He asked if I chewed gum and I replied yes and asked why he wanted to know. He told me that there were times when I was using the blackboard and talking simultaneously that he became confused. I told him that I fully understood his dilemma and would never chew gum when I was in the classroom again.

I taught 9th grade algebra in night school after my second year of teaching. I had a husband of one of my students come to my room looking for his wife at the start of my fifth session. I checked my record book and found that she had made only one appearance. That was on the first night.

House painting

I RECEIVED A PHONE call from a man who was in the process of selecting an individual to paint his three story farm house. My name came up as a potential candidate. I arranged an appointment time so that I could determine whether or not I was in agreement with the total desires of the homeowner.

The owner wanted two coats of paint applied to all exterior surfaces except for the rear screened in porch area. The first coat was to be oil based primer. The second to be latex water based paint. It was agreed that the owner would purchase the paint. I estimated that it would require 12 gallons of primer and 10 gallons of latex.

I gave the owner a final cost to paint his three story farm house, which he accepted. I told him that I would have to rent two sections of scaffolding and a 16 foot extension ladder before I started the project. This would be accomplished over the ensuing weekend so that work could start the following Monday.

I was blessed with excellent weather and did not have to scrape any of the siding prior to receiving the paint. I had one problem arise after I painted the front porch area. It rained soon after I had applied the second coat of paint. Any watered down paint that landed on the porch had to be sprayed with water. I was lucky that it was not a larger area that I had to repaint the next day.

The only difficulty I had with the entire job was to paint a circular window that was positioned high on the front side of the building to allow light to illuminate the attic. I first placed the ladder on a piece of carpeting that was placed on the metal roof that covered the front porch. I was convinced that the carpeting would prevent the ladder from slipping when I was forced to stand on the top rung of the ladder.

I took a pail with a small amount of paint up the ladder to paint the window because if I experienced difficulties I would not have to clean up a large amount of paint. I was about half done when I felt the ladder slip. The second slip forced me to grab the brush and paint and jump to the porch roof. I was lucky that no paint was spilled, the roof was not affected and I was unhurt. The ladder stopped sliding as soon as I was no longer on it.

I had another homeowner call me to see if I was interested in painting the windows on the second floor of his home. I gave him a price to paint 7 windows. He accepted my offer and expressed his desire to have me paint the windows any time during the next week. He gave me a gallon of paint for the windows. He offered to pay me in advance but I insisted that he wait until he came back from vacation.

My daily schedule was such that I did not have time to paint during the day. I decided to paint these 7 windows during the early evening.

The first evening I painted the four windows on the front of the house with no problems.

The next evening I finished the job, painting the remaining three windows. I was able to stand on the slight roof overhang. I used my ladder to gain access to the garage roof and then simply stepped directly onto the roof overhang. I had finished the painting and was ready to step on the garage roof when a large gust of wind blew the ladder off the roof. My dilemma was how I was going to get off the roof or how could I retrieve the fallen ladder.

At the very instant of the time when I was deciding what I should do, two teenage girls walked by the house and offered their assistance. I quickly told them to hand one end of the ladder to me while the other end dragged on the lawn. I was able to grab the end of the ladder and pull it up on to the roof. I told them how appreciative I was for their help.

Delivery Service

MANY BUSINESSES RELY ON a delivery service to guarantee that important documents and legal papers reach their clients in a timely manner.

It is important that interested recipients attain the documents far enough in advance to make them accountable.

I was told to fill up my vehicle with natural gas at the end of my business day. By doing it this way, the vehicles were always ready to travel.

The first few days on the job, I was able to learn knowledge that I did not know. I learned where the dead-end streets were, where the short cuts were between two locations, what alleys that I could use and where I could park illegally without receiving a ticket.

Each time I left the shop, I was given the object to be delivered, the address of the client receiving the object and his signature. I had to be given explicit directions as to where to locate the client for the balance of that day if he was not readily found.

I had to obtain the client's signature at any cost to complete the transaction.

I transported dental x-rays to other dentists involved with the same client in an advisory role. The same scenario was employed among doctors and their associates.

Occasionally I would have difficulty locating the client address. It might be one unit of several situated in the same apartment complex or a group of retail shops having the same address, only distinguished by a letter of the alphabet.

I picked up mail from several post offices at about the same time each month. Most of the envelopes were addressed to the same bank with several slightly different zip codes, a method used to balance the work load.

The company employed a dispatcher to aid those drivers who had a problem with a delivery. He also used large blown up maps to help the driver circle the area in question. The driver had a similar version in his vehicle that helped him in his search.

The dispatcher was also very helpful to a driver whose purpose was to deliver an object inside a gated community. I was to pick up an actor's trunk at one location and deliver it to another address the same day and backed up with two signatures.

Architects are fond of a delivery service as well. It enables them to approach several clients at the same moment in time to determine what they feel is the most productive use of the property.

It was sometimes necessary to consult the dispatcher when I had overshot my destination. He could inform me of the simplest way to backtrack to avoid any traffic jams. Sometimes he could employ his past association with the same professional to tell me exactly how to reach my destination.

Field stone dam

WHEN THE MAN CALLED me about a field stone dam, I was very interested in learning more about his idea. I arranged to meet with him the very next day at the dam site.

He led me to the dam's eventual location. He had already stacked 3-4 timbers (10″ x10″ x 10″) on top of each other. They were positioned in the dam's opening and were both bolted and nailed together.

He presented his idea to me. The sides of the dam would be built up with field stones of random sizes and then cemented in place. They would be assembled to achieve a wall thickness of 4 to 6 feet.

He owned property on both sides of the dam. One of the features that he wanted to achieve was to be able to operate his tractor not only across the dam's top edge but also to navigate the area beyond the dam.

His intent was to eventually position a large wooden water wheel below the dam. It would be designed to spin in only one direction and its movement (speed of spin) is determined by the flow of water.

This action was anticipated to produce D.C. electricity (direct current). It was considered to be the electrical source for the greenhouse which was already built as part of the landscape.

During construction of the dam, the water never stopped flowing. It provided us with a never ending source of water enabling us to produce a cement mixture whenever we needed one.

The construction of the side walls followed a distinct pattern. The layer of rocks comprising the dam's side walls (those facing the water) were built up first. Additional rocks would then be positioned as close to each other as possible. A dry mortar mixture was poured between the lower layers of rocks. It would take one day to become solid so that we could continue.

This osmotic feature enabled the water to mix with the dry mortar.

Our part of the project was completed within two weeks. The homeowner was confident that he could complete the balance of the dam on his own.

He planned to increase the height opening by adding additional 10" x 10"x 10" timbers to the ones already in position.

I happened to be in the neighborhood in late September and decided to see what state his dam was in. He was standing on a cat walk that was below the base of the dam. He had placed enough timbers in the opening to account for more than 6 feet in height.

He acknowledged that the water wheel was missing but would be in his possession within the next month. I was really surprised to see the spray of water coming over his head and extending horizontally for 5-6 feet before it cascaded to the ground.

I was not sure how he would harness the dam's energy to a level he could use.

Senior Perk

DURING THE FALL OF the school year, the senior class members held a candy sale. Any senior selling more than twenty dollars of candy was eligible to dine in a Hawaiian restaurant and attend a Broadway play.

Eighty seniors signed up to go on this excursion. I had to arrange transportation, select 8 chaperones and decide what date would be best.

My next mission was to determine, in advance, the total cost at the restaurant. My previous experience dealing with the restaurant personnel was confrontational and did not want a similar scene. The captains expected their tip to be given to them in cash and the coat check girl expressed the same demand. I was determined that I would not relish the same experience again.

I contacted the restaurant and told them my feelings. I would pay a set price for 90 meals and I would bring the check with me. Neither the captains nor the coat check girl were to ask for tips and should be alerted to this condition prior to our arrival.

The owners assured me that this matter would be taken care of. I was somewhat annoyed when I counted four more chaperones than I had allotted for. They had decided to invite themselves or were asked to partake by the officers of the senior class.

I requested a meeting with the owners, because I had already been approached by one of the waiters asking for his tip. They took me to the owner's office and I reminded him of our agreement that no employee was to request monies from my group. As we were leaving, the hat check girl asked for her tip. I used her phone to alert the owner as to what she was doing. He corrected the situation very quickly.

I made certain that the 2 bus drivers received their dessert early so that they could bring their buses close to the restaurant. The sidewalks were crowded with pedestrians when we exited the restaurant making it difficult to locate the two yellow school buses.

After sighting up and down the street for the buses, I bumped into the 2 drivers not 50 feet from the restaurant. I then had to lead the students to the buses, about 5 blocks away.

I did not have the time to do a head count and asked the theatre director to call the theatre to request a 15 minute delay before they started the performance.

We were later than anticipated and filed into our seats during a brief nudity scene (girl performers discarded their tops). I was uncertain whether or not this was part of the script.

The exit from the theatre to the waiting buses was smooth with the students very well behaved. I made a head count for each bus and realized we were missing two boys.

A new teacher, who lived in the city, suggested that we break into small groups to look for the two missing boys. I told this teacher that I had 78 students on the 2 buses and I was going home and not creating a ridiculous search for the missing boys. I suggested that he kindly remove his body from the school bus.

Once the students were safely deposited on their home turf, I decided to inform the principal that we came back home missing two seniors. He asked me what "we" should do about it.

I suggested that he call their parents, even though it was after eleven o'clock. I told him that it would be after midnight before I arrived home.

The following day the boys came to school late and were quickly sent to the principal's office. They informed him that they had told his assistant about their plans after the theatre and never told anyone else.

Two by water

I was sent with Matt to the lake to make some service calls. Besides the service calls, we were asked to distribute a pamphlet to every boat that might be interested in winter storage.

The pamphlet offered free pick-up in the fall as well as free delivery in the spring. The batteries would be charged also without any charge to the owner. Any necessary repairs would be given preferential treatment at a reduce rate.

Realizing that our venture was going to be time consuming and to avoid much personal contact with the boat owners, we left the boat shop around 7 a.m. We intended to start at the north end of the lake and circle the shoreline in a clockwise pattern. We were enjoying our successful endeavor as we progressed along the shoreline. Matt reminded me of our first service call. The owner had complained about his boat not starting early in the morning. Matt directed our boat towards the one requiring service. Before we could reach the shoreline, the owner's large German shepherd appeared from nowhere. He acted like I was his next meal. Matt put the engine in reverse and backed away from the shoreline hoping the dog would negate his desire to take a bite out of me.

The perspective was valid. He had been trained as an attack dog. Matt was familiar with the dog's reputation, but I was not. The dog made enough noise to awaken the residences of the house. They placed the dog in his dog house so that we could make the repairs to the engine. They apologized to us for their dog's behavior. We considered this incident number one.

We continued with the distribution of the pamphlets. We spotted the area of the shoreline where we would make our second repair. This house was built on an island and had its own bridge to cross to reach the house.

The owner was conversing with another gentleman when he turned and directed the gentleman to go to the house. When this happened, Matt stood up in our boat to announce that we had been sent by the boat shop owner to do some repairs on his boat.

When we docked our boat, the owner told us that he had anticipated a "hit" by the family. We fixed his engine and left to complete the distribution of the pamphlets. We considered this incident number two.

Jewelry guard

I WAS CALLED BY the detective agency to learn whether or not I could work as a guard from 10 am to 4 pm on this coming Sunday, the 16th of May for the jewelry store located on the second floor of the fashion mall.

I would be paid time and one half for my six hours of labor. I was expected to be present at the time of the opening as well as the closing of the store. I was supposed to be alert, to be moving around most of the time and not to enter the sales area unless my presence was requested.

They wanted me to stand just outside the sales area. If any of my friends came by to converse with me, I was expected to limit my conversation to a couple of minutes.

The chairs inside the store were to be used solely by the customers of the store.

For a change in scenery and the need to use the restroom, I would walk the two corridors needed to reach that area.

I tried to alter my path with each walk through the corridor to avoid the walk becoming monotonous. I developed illusions about chairs and benches that did not exist.

I became so convinced that they were real that I found myself attempting to sit down on one of them. I would catch myself before I actually sat down.

I visited the men's room and attempted to wake up by splashing cold water on my face. When I returned to my post I witnessed a robbery taking place in the costume jewelry store on the first floor directly below my post.

Two young white males were involved with the robbery. I was uncertain whether or not they had guns. One male did the robbery while the other one held the two bicycles used for their getaway.

They left the mall riding their bicycles. The store's employees did not make any noise during the actual robbery but did yell after the thieves had left the building. I was never asked by anyone as to what I might have seen. I was only aware of a robbery occurring once I saw them jump on their bicycles and exit the mall in haste.

I finished my assignment, never mentioning the robbery in the costume jewelry store on the first floor. I did not tell the guard service about the incident either but made it perfectly clear that I did not want the same assignment again.

Author

OUR HIGH SCHOOL OFFERED 3 programs to our students to determine if any of the three would be accepted as full time programs. We developed a questionnaire that would provide us with background information as to their likes and dislikes of each program.

The students answering the questionnaires wholeheartedly supported my horticulture class, going as far as suggesting I write a book about the subject. The encouragement by the students gave me the will to proceed. My initial thoughts were to put together just enough subject matter to cover one semester.

I developed an outline that I felt would cover the basic information about the field of horticulture. I elaborated on this outline to make the book saleable.

I knew no one in the publishing business so I needed some introductions from an outside source. My lawyer had a client who was a personal friend of several executives on the staff of Reston Publishing Company, a Prentice-Hall company. He sent me a copy of a letter he had sent to the executives at Reston. I received the copy and waited a full week before I attempted to call each of the four executives. I came up empty since none of the four were familiar with the letter or my name.

The very next day, Friday, I received two separate phone calls from 2 of the executives. I asked each one of them the same question. I called yesterday asking the same question of each of you. Both replied with the same statement. Yesterday is gone forever, today is what counts in the publishing field. I was told by one executive that I would be receiving a contract in the mail for my horticulture manuscript.

The expectations were to have me complete the manuscript in 9 months time. I realized that I was going to be busy gathering background information, organizing the material and eventually writing the text. Once the publishing company received the printed text, their personnel would proof read the text and make corrections. The text would then be placed in a galley proof with continuous text and with margins for marking corrections.

I was then sent the galley pages for me to proof read and make changes. The diagrams are noted by figure in the galley pages but do not appear until the text has been made into pages.

Once this procedure had been completed by the publishing staff, I again received their output. Each page was identical to the pages in the book, identifying where the figures would appear on each page.

The last phase was to place the diagrams in their respective locations on each of the pages. They sent the completed pages for my perusal to hopefully make any corrections before the book was printed. It was successful enough for the company to have a second printing. The book's title was "Horticulture-A Basic Awareness".

The next time I came in contact with an executive from Reston was for him to ask me if I had any new book titles. I informed him of a title "Lawn and Garden Construction". He asked that I send him an outline of the chapters in the book. If interested, the editor would send me a contract and request that it be ready for production in nine months time. The editorial process, mentioned earlier, was sped up to reach publication faster.

Once again the editorial staff was looking to publish a textbook entitled "Modern Carpentry" to compete with another text already in print and having the same name. I was told that I could employ several writers because of the enormity of the task. I asked several of my friends but they all tried to write a chapter of their choosing and were unsuccessful. I was finally the only one who was ready to undergo the task.

I started compiling information for the standard chapters. I added several chapters that were current and not any portion of the existing text books. It took one year of my part-time endeavors to deliver the text for the entire book and have it placed in galley pages.

I was asked for the figures that accompanied the text. I told the editor that the only art that was to support the text came from several hundred sources. The editor had not included any monies for art work in his budget. He asked me to seek out all the background art that I used in formulating the text. This task took about 9 months to gather all the figures. Finally, the book was published.

All 3 books have now been out of print. The only way left to purchase any of these three books is via the internet.

Door to door sales

I ANSWERED A NEWSPAPER ad seeking an individual that likes to sell door to door. I was unfamiliar with most of the proven sales techniques but would be willing to learn.

My first endeavor was to sell world book encyclopedia. I would initially learn all I could about World Book by studying the "B" book and reading the information in the sales manuals and pamphlets.

My superior would test me orally and then suggest that I accompany him on his next presentation to learn the proper techniques and how to close the sale.

He told me that most sales come from leads obtained from previous buyers. He suggested that I canvas my own neighborhood to ascertain whether or not there are clients close to my home. He offered to help me with my first presentation but I refused the help. I told him that I wanted to do the first two on my own.

My first attempt was with a family in my neighborhood that I was friendly with. They had three children and all three could benefit from having the encyclopedias in their home.

My presentation was "low- key" but I made the sale. They decided to pay for the set over time.

I made certain that I did not leave their house without the names of 6 other families that could benefit from having the set.

I had another presentation set-up for early afternoon and another at 6 p.m. My afternoon appointment went off without a hitch. I was very proud of myself and was convinced that I was "on a role".

For my third presentation, I had to drive about 25 miles, most of which was in unfamiliar areas of the state which I had never traversed before. I hoped that I would be in the vicinity of my eventual destination before it got dark and rainy.

I hit a pothole at the entrance to a bridge covering a stream. It was an engineering nightmare. On either end, the entrance to the bridge was straight and after 2 car lengths, the road curved to 90 degrees.

When I hit the pothole, the car's battery became dislodged causing it to hit the frame of the car, resulting in the car losing all its power. I opened the hood and used several pieces of wood to wedge the battery firmly in place and away from the car's frame.

The car came to rest in the middle of the bridge. I made a quick decision to push the car off the bridge so that it could be more readily seen by a passing motorist. I decided to wear one of the white laboratory coats to keep me drier than I would be without it. I also could be more easily seen.

The first car came by and both women felt that I was a doctor on a service call. I told them I worked in a medical laboratory. I asked them if I could use their car battery to jump start mine. They agreed with my request. I had jumper cables and as soon as I made contact with both batteries, I turned the ignition key and my car started right up.

I decided to go straight home and call the client to explain what happened.

Mirrored twins

MARIE, OUR KITCHEN HELPER, was given a job in the kitchen, but because of her age, 15, she could not be considered a full-time employee.

A great deal of time was spent in staff meetings learning as much as everyone could about their fellow staff members. I also wanted to know what were their likes and their dislikes.

I asked each staff member why they applied for a job at this particular camp and what did they intend to gain from their experiences while at camp.

Most were shy and timid and were reluctant to discuss their thoughts and philosophy about life with anybody.

I received some mind shattering answers from my staff as well as their universal thoughts about handicapped children. I was surprised that most of them agreed that they did not need to have anyone in society taking pity on them. This was also reinforced by the campers themselves. They wanted to be treated as normal even though they were very familiar with their physical problems.

Marie told us about her mother and twin brothers. Marie's mother was a devout catholic who had the misfortune of losing her new born son 4 days after birth.

This occurrence had a devastating effect on her plus the conviction of the doctors that she could not have any more children. She was determined in her quest for another baby and went to the shrine every day, rain or shine, to pray for another child.

Her prayers were answered and she became pregnant. When they were born, the gynecologist noted their respective positions in the uterus. The 2 were facing each other in the womb with the left side of one twin coming in close contact with the right side of the other twin.

Marie, her mother, my wife and I found a time to discuss the twin's behavior with each other and with other members of society. When they were in kindergarten and asked to draw a fall scene for their class project, the two drawings that the twins submitted were identical in many ways. One of the drawings, if superimposed on the other, would show marked similarities between the two.

Usually the educator in charge is convinced that both drawings were the product of the same student. The following week the teacher decided to have her students draw a winter scene. She would position the twins to guarantee that they could not copy one another (one in front of the class, the other in the rear).

The results were identical much to her amazement. I decided to give them a tour of the campsite. One twin was walking on my left, while the other was on my right. Without any conversation from anyone, the twins took off running simultaneously.

Several universities offered full scholarship to the twins, provided their parents allowed the college professors the opportunity to study them, both at home and in the classroom i.e.; the field of psychology and the study of the mind and behavior.

The mother listened to them before she ended up denying each of their proposals. She viewed the conception, the gestation period as well as the birth to be gifts from God.

To me, it is a phenomenon that we must accept and not be one that is questioned.

Attic conversion

A FRIEND OF MY clients called me requesting information about the feasibility of adding two bedrooms to the attic area of his single story ranch home.

I arranged to meet him on April 29th at 9 am. I arrived on time to discuss the possibilities of building two bedrooms in the attic.

The first item to be concerned with was the only access to the attic was an opening 15"x30" located in a bedroom closet. This was to be pushed up into the attic, while standing on a step ladder, to gain access to the attic to retrieve Christmas ornaments.

I examined the entire ceiling of the first floor as well as the arrangement of the furniture in the living area. A permanent staircase was out of the question.

He asked if a pull down staircase could be employed. I informed him that there was only one possible location where it would work. It would be installed so that the base of the steps would land in the center of the living room.

I suggested that if he were planning an attic pull down staircase, he should select the more expensive one. He asked me when I could install it. I told him that I would obtain prices from several lumber yards and select the best one. I would also give him a price for installation.

I informed him that for his children's safety, I would install a railing around the staircase, hinged on the left side to swing away from the door opening. It should completely encompass the stairwell with a locking clip. The door should swing away from the stationary portion of the railing.

It was two weeks before the two of us connected with each other. He was satisfied with the price of the pull down staircase, the protective railings, plus the necessary price for hardware to adequately do the job.

He asked me if I would give him suggestions as to how to frame out the remaining space.

He had several buddies that had promised him their help in making the attic area livable.

I never had a reply from him about his attic. I can only hope that he was able to make the attic area whole.

Renovations

RENOVATION IN THE CONSTRUCTION field suggests that the item(s) be restored in good conditions by repairing or remodeling them. Some craftsmen believe that renovation is easy to fix and should not be considered difficult.

Those who perform renovations are able to overlook some of the flaws and blemishes and be satisfied with an adequate solution.

I believe that the outcome of any restoration is satisfaction but never perfection. The craftsman has no insight as to how the wall was constructed, where the electrical lines run within the wall and whether the hall bathroom lavatory is equipped with a water hammer. These are three examples of areas within the home that sometimes need renovation.

In older homes, many were not constructed with a sliding glass door as a means of entering the rear yard. Because of a more active lifestyle, more homeowners are changing their exit door to a slider. It enables the mother of the family to enjoy several perks not afforded with a single-hung door such as: better visibility of the yard and easier access and ample size for the carrying of picnic items back and forth.

Families that have a combination storm and screen door for the rear exit face other problems. The hydraulic cylinder which is involved with the door closing mechanism is located in the center of the door on the hinged side.

By adjusting the clip (washer) on the accompanying metal rod, the speed at which the door closes varies with the position of the clip. If slid back against the cylinder, there is no hydraulic action allowing the door to close rapidly like a guillotine.

I personally can attest to this action. While adjusting the clip on the rod located on the left side of the door I entered the house with my right hand on the right side of the door frame. The door closed abruptly. The clip acted like a guillotine and removed the tip of my right middle finger. I have rerun this scenario a million times or more and cannot explain how the fingertip on my right hand was removed by the action of the hydraulic door closer located on the left side of the door frame. It spun my body 180 degrees so that I ended up facing the yard.

The damaged finger was reduced in size to be the same length as my index finger. The doctor was unable to sew the tip back on the cut. He told me that once a person reaches an age of 35, the two portions will not rejoin.

I received a phone call from a former client regarding the action I would suggest to correct a serious problem. He was planning to close on his home on Friday and had turned his baseboard heating system off the previous week. He was convinced that every aspect of his home was presentable to the eventual buyer.

He did not anticipate any problems until he visited his home and discovered that his hot water system had erupted creating damage not only to the water pipes but also to the drywall. The oak wood flooring in the living room was pushed upwards into the shape of a "V" with the tongue and groove components still intact. He asked me if I had any ideas as to how to replace the living room floor so that the sale of the house could commence as planned.

He could have a plumber repair the damaged pipes and someone else to do the drywall repair.

I told him the "quick fix" was to face nail each piece of flooring throughout the living room floor back to its original level state. The spaces were to be filled in with epoxy.

The floor would be receptive to any type of flooring. I told the client to have the new buyer, within a day or so, make a selection and have it installed by Thursday a.m. They would have to be responsible for the carpet clean-up.

The owner thanked me for my input realizing my suggestion was the best solution for all concerned. All the sub-contractors completed their work on time and the new owner was extremely elated to have brand new carpeting in the living room.

Christmas tree sales

FOR MANY, SELLING CHRISTMAS trees has a certain mystique about it. Many view the job as being ideal because the tree sells itself.

I spent one of my vacations in an outdoor parking lot selling trees to homeowners. Most of them wanted to buy the tree of their wishes for a reduced price. Trees were available for sale the day after Thanksgiving.

Time was spent constructing wooden tree stands with the hope that they would sell easier and faster with a tree stand than without one.

Little did I know that the trees were cut down in late August or early September because of the fear that if they waited any longer, the foresters would encounter deep snow drifts and would be unable to transport the trees.

They relied on the temperatures remaining cold to stall the respiration rate of the trees. Unless the weather turned warm, the trees would be alright. Once taken indoors, where the temperature was 70 degrees or warmer, the tree became dried out quickly. To slow this process down, the base of the tree trunk was immersed in a container of water. This action enables the trunk to imbibe or soak up the water to replenish that lost by evaporation through the tree's needles.

Water must be added to the container daily. If this addition of water is either ignored or forgotten over several days a fire might be a strong possibility.

I sold trees at a well known nursery during another vacation period. I was the only salesman in the entire rest of the nursery. The nursery employees were attending a Christmas party held in a barn nearby.

My main function that afternoon was to sell the remaining trees to a pair of buyers who had been given a quote of $8.00 per tree. I counted the remaining trees and came up with a total of 32 trees. There was no charge for 2 of the trees because they were considered unsatisfactory. My boss wanted $240 dollars for the trees, money to be paid in cash and removal of all 32 trees and any vestiges of tree branches. He expected that they would arrive about 3 p.m.

I was expected to collect the money at the onset of their arrival and take it immediately to my boss.

With that accomplished, I returned to the site of the trees to make certain that the other aspects of the deal were complied with.

I asked the two enterprising males what they intended to do with so many trees. They told me that they would select a different area of the city having a predominance of high risers, park their truck close by, using the truck as the sales platform.

The residents in those buildings are always cramped for space and have no area in their unit to allow for a Christmas tree to be erected.

The children living in one of the units accepts the belief that Santa brought the tree with him. Many residents set the tree up on Christmas Eve and take it down the following evening.

Adult Education

I LEARNED ABOUT AN instructor needed to supervise separate classes of physical fitness for females, 2 nights per week and the same schedule for males.

Each session was 2 hours in duration. The females first hour was spent doing physical fitness exercises. Two of the women preferred yoga to calisthenics and spent their time going through the various positions they were familiar with.

Each person was charged $15 for the ten sessions and the monies were expected to be paid during the first session. Two of the women did not have their money on their first session or on their second session. Each of those two women asked me to allow them to participate without paying for the class, suggesting tight family finances. One paid and the other did not.

I reminded them that I was required to take attendance at each session and would be unable to falsify my records. The school district, in turn, was required to submit their records to the state education department handling the adult education classes for reimbursement purposes.

At the final session, I distributed a questionnaire to each woman participant requesting their suggestions to improve the program. They were happy with the way the program was run and their only comments were to extend the time spent each session (up to 3 hours in duration) and extend the sessions to either 15 or 20.

The men's first hour was spent playing 4 on 4 basketball with the first team to reach 11 was the winner and would then play the next 4 players on the sidelines.

The second hour involved a very spirited game of volleyball for both the woman and men. Those not involved in playing volleyball spent their second hour on the mats doing calisthenics. Their questionnaire reiterated the comments made by the women.

I spent two school years in this capacity and was very optimistic that it would continue for several years. I was wrong. A teacher from the school district expressed a desire to supervise the program leaving me without the job.

I heard about teachers needed at the local BOCES facility. Each teacher had to submit a field of expertise outlining the topics that would be covered for each subject heading during a 10 week schedule.

I submitted 3 subjects that I felt comfortable presenting to interested adults. The first was a course in basic horticulture. It would cover the following topics:

1. Cultural requirements of plants
2. Watering plants
3. Garden practices
4. Planting
5. Lawns and their care
6. The vegetable garden
7. House plants
8. Insect and disease control
9. Specimen trees and shrubs
10. Ground covers and mulches

The second topic that I would offer would be entitled lawn and Garden Construction and would cover the following topics:

1. Steps involved in concrete work
2. Lumber and related wood products
3. Estimation of materials
4. Planning the project
5. Decks
6. Patios

7. Garden steps
8. Walks and garden paths
9. Fences
10. Walls

The third topic was masonry and it covered the following topics:

1. Tools needed
2. Materials involved-sand cement, pebbles
3. Forms-concrete-slope
4. Blocks-how to use with/without mortar
5. Bricks-patio, walk ways, steps
6. Field stone-walkways, walls steps
7. Techniques- dry mix, wall pargue
8. Footings-for wall posts, piers
9. Stucco-dry wall, texturing, plaster
10. Planters-drainage

I have taught all three of these topics to adults and they were well received.

Magazine articles

I RESPONDED TO A newspaper advertisement seeking someone who could discuss the necessary steps needed to follow the construction of a fieldstone wall with a set of steps incorporated into the wall.

I called to make an appointment for next Friday at 9 a.m. the timing of my visit was "perfect" according to my client. He mentioned several times how lucky he was to have me honor his phone call in such a punctual manner.

When I arrived at his house, he showed me the rock wall around the base of a tree. I explained to him that to make the existing rock wall higher and construct a set of steps as part of the wall would be almost fruitless.

He finally stated his intentions. He had signed a contract with International Harvester to develop a magazine with a gardening theme. He was somewhat behind schedule and would welcome any help I could give him. Once he discovered that I have authored two books, one on horticulture, the second on garden construction, he first wanted to know the chapter headings in both books.

He was so enthused for finding me and my background of usable knowledge plus my ability to edit copy. The more I did for this producer, the more he requested I do. In the beginning I was supposed to write and edit only two articles but the final assignment was much greater.

He produced several magazines about car care. They ranged in scope from a basic tune-up to complete engine overhaul. I had enough background knowledge about car repair that I was a great asset to enable him to produce an excellent magazine.

He was producer of a camera magazine sponsored by Canon. The magazine was all ready to be sent to the printers when someone noted that Canon was spelled with one (1) "n" in the middle of its name. I felt sad for my boss, the producer, and thankful that I was not involved with any phase of that magazine.

Corrections were made before it was returned to the printer. The cost to reprint the magazine was almost twice its original expense.

The producer continued to employ craftsmen to erect various structures on his property. He would take a substantial number of photographs depicting the actual steps involved in the construction of each structure.

Sometimes, he would rely on my construction knowledge to explain to the reader of the magazine in a series of steps, how an item in the landscape was assembled. For example: he hired two carpenters to build a gazebo on his property. He took a multitude of photographs as they assembled the gazebo, always with the conviction that I would know exactly how it went together, thus eliminating the guesswork in the reader's mind.

There are commercial establishments who provide interested authors and magazine producers with suitable photographs that they may find useful and appropriate to embellish their articles.

One fact that you must understand, if you author an article, you will be paid only once for your article. If you write a book, you will receive royalties for each book sold. The magazine article can be reprinted over and over again with no additional payment to the author.

Caboose

My neighbor owns a country western shop in our community. He called me a few nights ago to ask me if I would be interested in preparing the landscape in front of his store to accept a caboose from the railroad.

His first request was to align the side entrance of the caboose with his present building, an old country school house where he personally sold the adult clothing.

His only concern was that space be provided between the two to accommodate a ramp large enough for two way foot traffic.

I accepted the challenge and was told by the owner that the exact measurements could be obtained by checking other cabooses in the surrounding towns. I needed to learn the length of the rails, the distance between them as well as the space between the railroad ties used as support for the two rails. The space between the railroad ties was filled with gravel to the height level with the height of the ties.

We went ahead as planned staking out the various components of the railroad track. The rails, roughly 20 feet in length, had to be installed perfectly level.

The railroad ties were then positioned in their location, making certain that the distance was the same between the rails and that they over hung the rail by the same component.

Once every facet was checked, leveled and rechecked in every direction, we were ready for the caboose.

The last lap of the caboose's journey was to climb up a very steep incline, preceded by a crane on a flatbed trailer.

The owner worried about the length of time the installation would take. Surprisingly, the caboose was placed on the rails in less than 30 minutes. Eight metal chocks were installed at the ends of the rails to stop any forward or backward movement.